T0296198

TEACHING ASTRONOMY
IN SCHOOLS

TEACHING ASTRONOMY
IN SCHOOLS

BY

ERNEST AGAR BEET
B.Sc., F.R.A.S.

SECOND EDITION

CAMBRIDGE
AT THE UNIVERSITY PRESS
1962

CAMBRIDGE
UNIVERSITY PRESS

University Printing House, Cambridge CB2 8BS, United Kingdom

Cambridge University Press is part of the University of Cambridge.

It furthers the University's mission by disseminating knowledge in the pursuit of education, learning and research at the highest international levels of excellence.

www.cambridge.org
Information on this title: www.cambridge.org/9781316509791

© Cambridge University Press 1962

First edition 1956
Second edition 1962
First paperback edition 2015

A catalogue record for this publication is available from the British Library

ISBN 978-1-316-50979-1 Paperback

PREFACE

The origin of this little book can probably be traced to a talk that I gave at a meeting of the British Astronomical Association on my efforts to teach some astronomy in a school. Following this I got into correspondence with a number of other teachers and prepared an article entitled 'Astronomy for Schools' for *School Science Review*.* This was published in 1949, and since then I have had a number of opportunities of lecturing to teachers on this subject, and discussing its problems with them. In the light of these discussions and with the lapse of time the article required amplification; here it is, and I hope that fellow-teachers will find it useful.

NAUTICAL COLLEGE,
PANGBOURNE   E. A. B.
5 May 1955

PREFACE TO SECOND EDITION

In the interests of economy very few changes have been made in chapters I–IV, though some addenda to them will be found on page xii. Chapter V has been reset to incorporate many changes, both by addition and withdrawal, in the teaching aids available. There are also substitutions and additions in the Bibliography.

December 1961   E. A. B.

* The original talk appears in the *Journal of the British Astronomical Association*, Vol. LVI, No. 3, 1946; the *School Science Review* article is in Nos. 111 and 112, 1949.

CONTENTS

ABBREVIATIONS

The following abbreviations are used:

B.A.A.	British Astronomical Association
I.A.A.M.	Incorporated Association of Assistant Masters
J. Brit. Astr. Ass.	*Journal of the British Astronomical Association*
R.A.S.	Royal Astronomical Society
S.M.A.	Science Masters Association
Sch. Sci. Rev.	*School Science Review*
Whitaker	*Whitaker's Almanack*

NOTE

The bold superscript letters in the text indicate
references to the Addenda on p. xii

INTRODUCTION

Astronomy is certainly one of the oldest of sciences, and in this country its teaching at university level seems to have been recognized for a long time, for there has been an Oxford professorship since 1619. What has happened to it at school level? Little science was taught at all until the middle of the last century, and the appointment of J. M. Wilson (afterwards Headmaster of Clifton) as a science master at Rugby probably marks the beginning of science teaching in the sense that we use the expression today.* Rugby's first laboratory came in 1860 and its observatory a few years later. Of the latter the Devonshire Commission of 1875 reported that it 'is now playing an important part in the science education of the school'. The same report refers to telescopes at Eton, Rossall and Clifton. Of early textbooks there was one by Lockyer in 1868, and another by R. S. Ball in 1877.† Quoting from the Editor's Preface to the latter: '. . . it is very generally acknowledged by those who are practically engaged in Education . . . that there is still a want of Books adapted for school purposes upon several important branches of Science'. And in the Author's Preface: 'The present volume is intended for those pupils of the higher classes in schools, who, having some elementary knowledge of mathematics, desire to gain some information about Astronomy.' This certainly suggests that astronomy had a place in modern teaching, and yet by the turn of the century it had virtually disappeared.

* H. Armstrong, 'Experimental Physics Teaching', *Sch. Sci. Rev.*, No. 104 (1946).
† Sir R. S. Ball, *Astronomy* (Longmans Green, 1877).

Possibly this was a stage in the development of teaching methods. Laboratories became available, and the natural consequence was to use them; H. E. Armstrong was a powerful figure in the teaching world in the nineties, and he favoured the heuristic method for which chemistry lends itself so well. Thus chemistry held the field (for boys), supported by physics and with the rest nowhere.

There are, and probably always have been, schoolmasters (and no doubt schoolmistresses too) who are also enthusiastic astronomers, and have not failed, by hook or by crook, to pass some knowledge on to their pupils. Most of them ploughed lonely furrows; they did not make their work known and they received scant recognition. An exception was E. O. Tancock, who taught astronomy at Giggleswick and in 1913 published his lessons as a book* which remained the standard school text for years. While unknown teachers were keeping the subject alive it received official recognition from another quarter.† In a parliamentary report published in 1917, three recommendations are made: that in Preparatory Schools, at age about 12, 'physiography' should be taught and should include 'simpler astronomical phenomena, which in the hands of a good teacher may be made an excellent training in reasoning and observation';* that outlines of cosmical physics be taken with the non-science sixth; and that there should be short courses in astronomy for teachers. After the first world war the General Science idea began to bear fruit; the monopoly of chemistry and physics, already supplemented by some biology, was shaken and other branches of science, including astronomy, were considered by syllabus makers. The General Science Sub-Committee of the Science Masters Association included

* E. O. Tancock, *Elements of Descriptive Astronomy* (Oxford, 1913).
† *Natural Science in Education* (H.M.S.O., 1917).

astronomy in their second report issued in 1938, and some examining bodies included it as an option in the 'two-credit' General Science papers of the School Certificate—a practice that survives in the present General Certificate of Education.

Objections can be made to the introduction of astronomy into the school course, and the reader must decide on their validity. The curriculum is already too crowded—with the lapse of time the child's plate has become loaded with less and less of more and more. Has not the time come when for every new subject added an old one must be taken away, and what can we afford to take away? If the aim of science teaching is to give that attitude of mind called the scientific method, does it matter what sections of the subject are used for the purpose? Surely we should choose something that is convenient to teach, something in which the process of experiment-observation-inference can be most easily applied; astronomy is not such a subject. If the aim is to stuff the child with 'useful' facts, then astronomy is not 'useful' to most citizens—he is unlikely to make money out of it or even to earn his living, for professional astronomers are few. If the aim is just to get as many passes as possible in the General Certificate, then why waste time on something that provides only two questions, optional at that, in a very wide examination paper. Astronomy must surely be an expensive subject, and most schools are short of money. Another objection that I have heard more than once, from outside the teaching profession, from amateurs who love astronomy and say that it is 'much too precious to spoil by associating it with school'! Presumably readers of this book are already converted or they would not be reading it, but nevertheless the question of 'why teach astronomy' must now be considered.

ADDENDA

a *Page x: Preparatory Schools.* Many I.A.P.S. schools have recently adopted the loan service of scientific books and equipment operated by the Esso Petroleum Company. One of these termly 'units' is on astronomy.

b *Page 10: the syllabus.* Astronautics is a science in its own right and is far from being a hundred per cent astronomy, but some reference to it must be made with any present-day class. The minimum syllabus should therefore include an elementary description of the *astronomical* achievements of space research, and artificial satellites should be added to Extension 5. Radio astronomy calls for mention in Extensions 1 and 4.

c *Page 23: star maps.* The Science Museum postcards Nos. 372–375 will be found useful for distribution to pupils, the first two before Christmas and the other two after it. (Address on page 54.)

d *Page 38: simple telescopes.* A simple telescope of the kind described, giving a magnification of 25–30, is now available commercially from Charles Frank Ltd, Saltmarket, Glasgow. The lenses can be purchased separately.

e *Page 46: reflecting telescopes.* An inexpensive 'kit' for making a wooden 4-inch reflector can be obtained from Ottway & Co., Orion Works, Ealing, London, W. 13. Much care is needed in its assembly and if the pupils are doing it there must be a skilled handicraft master in the background. A kit for a standard type of 6 inch is made by Stanley & Co., New Eltham, London, S.E. 9, and a number of manufacturers will supply finished components for telescope builders.

CHAPTER I

WHY, WHEN, AND HOW?

WHY?

A professor of psychology once wrote: 'There is no need to dogmatize about the educational value of astronomy. As with any other subject, its value can be destroyed by bad teaching. But its *potential* value can stand comparison with that of any subject whatever. In young children an interest in the stars must be almost universally spontaneous, and this gives astronomy a flying start.'* There have been a number of investigations into the interests of children, three of which, all with a bearing on the present topic, have come to the notice of the writer. The method was similar in all three, that of getting the children to ask questions, and a general conclusion is that of the scientific interest shown the astronomical proportion was in the region of 15 per cent. The specific aims of the three investigators were different, as was their material.

Ball† used 30 classes from suburban elementary schools, and in his paper gives a detailed analysis of the work of seven of them, four of age 11 and three younger. Among questions classed as 'general' there were 11 of an astronomical nature; of 475 questions classed as physical phenomena, 55 related to the heavenly bodies, and another 24 occurred in the section on 'the world'. Many of the actual questions asked by the children are quoted in the paper and form a useful guide in planning a syllabus for the

* G. P. Meredith, 'Astronomy in Education', *J. Educ.*, Dec. 1949.
† H. R. V. Ball, 'Children's Interest and Experience in Relation to Science', *Sch. Sci. Rev.* No. 67 (1936).

youngest age-group. These results certainly support Meredith's 'flying start'. If we are to make use of the children's interests in planning a science syllabus, and if, therefore, we are to gain their maximum co-operation in making it work, then astronomy must surely have its place.

Richards* worked in one large urban boys' secondary school, so here we have a higher age group, and his aim was primarily to see how interests changed from 11 to 14+. The questions asked, a few of which are quoted in his paper, were generally similar in nature and distribution to those recorded by Ball, though with a slight drop in astronomy in the second year only. He concluded that at the beginning of this age range the questions were largely stimulated by curiosity—'Why?'—but at 14 utility—'What for?'—was a major thought. For astronomy the wonder stage is more useful, and this may be taken as an indication to teach it young.

Rallison† used 1659 boys and 1855 girls, about equally divided between town and village schools, to examine the effect of environment and sex upon their interests over the age range 11–14. He found that 78 per cent of the boys' questions were scientific, but only 48 per cent of the girls', and that environment had a greater effect on boys than on girls, town boys being more scientifically minded than village boys. Astronomy followed the same trends, with the additional one that whereas the town boy retained his interest over the age range, the village boy did not. This is a little surprising as it is only the latter who has a chance to see the sky properly, but we must bear in mind that Rallison's work was probably done during the war years, when, owing

* T. T. Richards, 'Pupils' Interests and the Teaching of Science', *Sch. Sci. Rev.* No. 86 (1940).

† R. Rallison, 'Scientific Interests of Children', *Discovery*, Vol. VII, p. 51 (1946).

to the blackout, town dwellers discovered the stars for the first time and there was a phenomenal rise in the sale of astronomical books.

One of the aims of education is to cultivate the right use of leisure. The stars are available for all to see, for nothing. Anyone who has to be out in the dark, as the fire watchers discovered during the war, can gain much from 'making friends with the stars'. Nature study, 'making friends with the countryside', is commonly taught, so why not its night-time equivalent? Then there are the impacts of astronomy on daily life, so commonly misunderstood—time, tides, the harvest moon and so on. Astronomy has its place in litera-ture, history, classical legend; the origins of the very names that we use are almost lost in antiquity. Science students are often accused of being uncultured. It is a doubtful accusation (and the scientific ignorance of the accusers is often lamentable!); science students need not be uncultured and astronomy in particular is, or can be, a cultural subject. Astronomy has already been referred to as an excellent training in reasoning and observation. A few years ago I heard a professor of physics address a gathering of local science teachers on what he would like his students to have done before they left school. One of the things that he said was 'give them some astronomy, material that they cannot touch, *and teach them to think*.' The italics are mine.

WHEN?

The question of the age most suitable for this subject has already been mentioned, 12 + in one case, and early in the 11–14 age range in the second reference. When I was teaching in a school where science began at 11 I did some astronomy in about the fifth term, just before a change-over from elementary general science to formal physics

3

and chemistry. I was quite satisfied with the boys' attitude to the subject at this stage, just about their thirteenth birthday. In my present establishment the boys do not join until they are 13, and such astronomy as I am able to do comes in their first term. I know of other schools taking it at 13, and also of one where it forms a major part of the first year physics course at 11 +. In addition I do some at the sixth form stage, though here the purpose and approach are different. In the thirds the aim is the stimulation of wonder; in the sixth it is to apply the physics and mathematics already done. One group has a section on gravitation in its advanced level physics, and another group is about to leave school as potential officers in the Merchant Navy; each group is treated according to its needs.

Other schools touch on astronomy in their sixth forms, and I heard of one where they ran separate courses for the science and non-science sections. In the middle forms, where I and many other teachers are still subject, in spite of the abolition(?) of the School Certificate, to the demands of external examinations, I find a ready interest in any astronomical topics that turn up. The subject is acceptable at any age, and I feel certain that it could profitably be included in the extra year arising from the Education Act—it must not be regarded as 'for grammar schools only'. And in the exam-ridden grammar schools 'There is not sufficient recognition of what can be done within the G.C.E. framework. There is, in fact, endless opportunity for unexamined work. . . . I would beg the schools to take every opportunity of using the freedom and elbow room granted by the release from what were so often called the shackles of the old School Certificate.'* It is easier for a

* J. F. Wolfenden, to a meeting of Convocation in the University of London, Oct. 1954.

4

Vice-Chancellor to talk than for an assistant teacher to act, but this outlook should at least be discussed with the headmaster.

There is a trend of thought that astronomy should not be in the curriculum at all, but should be a recognized out-of-school activity. Possibly this arises from the idea, now obsolete it is to be hoped, that anything done in school hours is necessarily hateful. My own feeling is that we cannot have one without the other. To quote the psychologist again: 'But it would be a pity if school astronomy were conceived solely as a branch of the regular curriculum. We all know what can happen to Shakespeare as a school "subject", and the same can happen to astronomy. A fostering of a lively interest by day will drive many a lad to spend half-hours in his own back garden by night, gaining that irreplaceable experience of directly reaching out to the realities of the ever-changing heavens.' Out-of-school, only the enthusiasts will participate, and no doubt better work will be done. But if this is all, how many possible enthusiasts never discover it? In school hours the possible joys of a new hobby can be offered to all, and there will be no lack of those who want to go on in their own time. Another weakness of the 'out-of-school only' system is the difficulty of laying a secure foundation. It is not easy to give formal instruction in the informality of a society meeting, and yet without it much muddled thinking can develop later on. My plea is for some school time, even only a little, to be given to a foundation laying for all pupils, and then a spare-time follow-up for those who will.

HOW?

At this point we are mainly concerned with how to fit in astronomy with the existing order. Of course it may be

5

easy, as in the grammar school where the 11-year-olds did it regularly for a year, or in another where astronomy in its own right had two periods a week for two terms in the fourth. Not all teachers will be able to set aside a block of time like this, but will have to work astronomy in with something else.

Consider first the geographer. As a part of his own subject he must deal with the Earth as a planet; good school atlases contain diagrams of planetary orbits, eclipses and so on, or even a star map. There is no need to labour the point; if this work is taken fully instead of being hurried out of the way as not being real geography, and if some time be given to the questions that a live class will undoubtedly ask, then here is an astronomy course ready made.

What of elementary mathematics? Astronomy cannot replace the teaching of mathematical rules and principles, but astronomical examples can be used as exercises: 'Plot to any convenient size the position of the Moon in its orbit when it is six days old, and show on the diagram the direction of the Sun's rays that shine on the Earth. Find by means of practical geometry, the shape of the Moon when it is six days old'; 'Assuming that the diameter of the Moon is 2000 miles, and that its distance from the Earth is a quarter of a million miles, calculate, as a fraction of a degree, the angle subtended at the Earth by a diameter of the Moon.' These questions are quoted almost at random from a book by P. F. Burns (5),* in which a course of astronomy is wrapped around the practical geometry that, to a greater or less degree, must be done in all schools. As an exercise with ruler and compass, working out the phase of the Moon is surely as useful as placing a polygon neatly inside a circle, and the exercise is sure to be followed by questions about the moon's surface.

* The numbers in parentheses refer to the Bibliography, pp. 64–69.

When I was teaching in the fifth term of a four-year course I took three physics periods a week for a combined syllabus in astronomy and light, the first round in this section of physics. I emphasize the word *combined*. We began with day and night, leading to rectilinear propagation, which through the usual laboratory work went on to shadows, and hence via eclipses back to astronomy again. When we got to position on the Earth's surface the class were shown a sextant; 'what is the mirror for?', and so over to the laboratory benches once more for experiments on the laws of reflexion. As an illustration of the method that is enough. Lenses linked up with astronomical instruments, dispersion with the constitution of the Sun, candle power with magnitudes and stellar distances. Of course, this was only a fifth term and the physics was not taken to the full School Certificate standard; that was done by itself in a later year. My general impression was that both subjects gained through being combined. In the 1938 General Science report of the S.M.A. astronomy was placed with physics in the first and second years.

Another example of a combined course is that taught by C. H. Dobinson (now a Professor of Education) in the 1920's, astronomy and geology, and set out in an attractive textbook (6) to which further reference will be made later. The same arrangement occurs in Dr Parson's *Everyday Science*, one of the first of the 'general science' textbooks published in 1929.

Astronomy for the sixth form does not really present a problem; they can assimilate lectures, and read for themselves more effectively, thus covering more ground in less time. Our subject can also be taught as a break from routine without giving a set period to it. With my present third form I make no bones about it—I just say 'no physics to-day; I want to do some astronomy'—and I really think that the change does more good than harm to the physics itself. The days concerned are chosen when there is some current astronomical phenomenon to discuss, or to serve as a starting point: 'have you noticed the evening star yet?' can start an effective lesson on the planets in general. We often hear teachers complaining about having a few weeks left after the examination, when a return to normal studies is somewhat of an anti-climax. A short course of astronomy will keep the class quiet and profitably occupied; I have tried it often, as has one of my colleagues, and it works well. Thus there are various ways of fitting in the new subject, and readers must choose in the light of the needs of their own schools.

CHAPTER II

IN THE CLASSROOM

There is nothing very original about the lists that follow;
a number of syllabuses have been published before. One
will be found in the current *General Science Report* of the
S.M.A. (4); at least two occur in the regulations for the
General Certificate ordinary level general science (Northern
Universities J.M.B. and Cambridge Local); there is some
guidance on this in Brown's *Teaching Science* (2). There is
much agreement between them, and what follows is roughly
the same; it is included for the sake of completeness, as a
book on teaching an unusual subject must contain some
reference on what to teach. Astronomers have criticized
this kind of syllabus as being nineteenth-century in outlook.
Over the last fifty years the centre of gravity of astronomy
has changed; modern professional research is largely directed
towards the remoter parts of the universe. The astronomer
feels that there should be some reference to his work in
modern education, and so there should, but in the sixth form.
The teacher knows that with the youngsters a start must be
made from their own experience, and that begins near home.
Thus a nineteenth-century syllabus appears once more.

The first section gives a basic minimum that should be
offered to every young person, though not necessarily in the
order in which it is printed. That order is quite suitable
when astronomy appears on the timetable; when it is to be
worked in with something else it will have to be modified,
but that does not matter as long as all the topics are duly
covered.

A basic minimum (mainly for lower forms)

Shape of the Earth.

Rotation of the Earth; the Plough and the Pole Star; circumpolar constellations.

Revolution of the Earth; seasonal constellations (Orion, Leo, Cygnus and Pegasus are enough to begin with); the Zodiac (not in detail); the ecliptic.

Tilt of the axis; 'plane of the ecliptic'; seasons.

The Moon; apparent motion; phases; eclipses (briefly); mention that tides are associated with the Moon; surface (briefly).

Other planets; apparent motion; distinction between planets and stars; the Sun (very briefly) as a typical star.

Suitable textbooks: Beet (7), Burns (5), Tancock (8).

Now follow some additional topics arranged under headings derived from the preceding chapter. Except for the sixth form, it is not intended that this work should be in succession to the basic syllabus. The teacher, having answered the question 'how?' according to local circumstances, should take the appropriate extension, combine it with the basic, and then write out his own teaching syllabus.[b]

Extension 1. Astronomy in its own right

Further constellations; the planisphere; classical legends.

The Moon's surface in more detail.

History and physical condition of the planets.

Comets and meteors.

The Sun; spots; prominences; corona; temperature.

The Stars; nature; distance; magnitude; colour.

Double and variable stars; novae.

Clusters and nebulae (briefly).

Nature of the Milky Way.

Amplify revolution and the ecliptic.

Latitude and Longitude; time (sidereal, solar, mean, zone); R.A. and declination.

Textbooks for teacher or pupils: Beet (7) or Tancock (8).

Extension 2. *Geography*

Time (sidereal, solar, mean, zone)

Latitude and Longitude; simple ideas about navigation.

Tides; springs and neaps; mention the effect of land masses; the establishment of the port.

Textbook for pupils: a good school atlas.

Extension 3. *Elementary mathematics*

Geometrical exercises on the basic course.

Further work on planetary motion; relative velocity; synodic periods.

Gravitation and the inverse square law.

Problems on time and the calendar.

Textbook: Burns (5).

Extension 4. *Elementary physics*

Some of the basic work can be incorporated with light (see page 7).

Shadows and eclipses to be taken more fully.

Astronomical telescopes.

Stellar magnitudes; variable stars.

Simple treatment of applications of the spectroscope.

Textbook: Beet (7).

Extension 5. *Science Sixth*

Gravitation; determination of G; Kepler's and Newton's Laws; problems on planetary motion; masses of planets and satellites.

Doppler's principle; line-of-sight velocities; extra-galactic nebulae; expansion of the universe.

History of astronomical discovery.

Astronomical measurements.

Textbooks for teachers: Spencer Jones (9), Barlow and Bryan (11), Part I of Nightingale (10), Doig (32).

Astronomy seems to lend itself, more than the other science subjects, to lecturing; it is the kind of subject in which the teacher is apt to stand up and talk. Is this because, unlike most of them, the material is not in the classroom, or is it because the other subjects have been in school long enough for a teaching technique to have been developed? Whatever the reason, lecturing at 11 + will not do. The lesson must be a co-operative affair, and Dobinson, in his book *Earth and Sky* (6), showed that it could be. Here is the beginning of one of his chapters:

THE EARTH'S TILT

In the last chapter we were recording mainly the results of the work of modern astronomers. Let us return to what we can see with our own eyes, and let us put ourselves in the police dock and give ourselves a little cross-examination.

Questions

1. Are the stars which we see at different times of the year always the same ones?

2. Are there any stars which we see at all times of the year?

3. Do people in Australia see exactly the same stars as we do?

4. Do people in Australia see any of the same stars as we do? If so, in what part of the sky would they be for us?

Let us consider the diagram below.

Then follows discussion and explanation of the questions. This is how a lesson should proceed; right at the outset the pupils must draw on their own experience and think, though in the examples quoted the experience would probably be less clear than in 'Have you ever seen the moon in daytime? If so, at what time of day?' Dobinson's chapters are interspersed with cross-examinations; at no time can the class sit back and pretend to listen. At the end of the chapter (or lesson) there are questions of the normal type and suggestions for things to do: make this, read that, go and look at something else, in fact the beginning of the out-of-school follow-up referred to in the previous chapter. Of course the teacher can talk to his class for a bit; he can, and must, show them pictures; but he must keep their minds working at the same time. Experienced teachers know all this, but at the risk of boring them it has been set down because in this subject it is all too easy to forget it and just talk.

Laboratory practical work as normally understood, does not exist, and much of the outdoor work (Chapter III) falls outside school hours. We must therefore consider what demonstrations and exercises can take its place.

FUNDAMENTAL MOTIONS

The rotation of the Earth and its consequences should not present very much difficulty. A geography globe is useful for demonstration, but an ordinary ball will do. A particularly dull child may stand up and turn around, and thus see for himself that objects around the walls cross his view and then vanish, while the ceiling is always in sight. The consequences of the annual motion around the Sun is more difficult to get over. Regard the walls as constellations and move the ball (Earth) around a lamp (Sun); blacking out the room is not essential but makes things more effective. Get the class to tell you what parts of the room are visible from the night half of the ball, what part of the room is obstructed by the lamp, and what part is always visible; this should lead to the distinction between circumpolar and seasonal stars, the path of the Sun through the Zodiac, why the stars rise earlier each night. An alternative for young or dull pupils (one master told me that in his school astronomy was reserved for weak sets only!) is to bring two out in front as Sun and Earth and get 'Earth' to announce what he sees as he goes around: 'Sun's head's in front of the blackboard—now it's in front of the bookcase' and so on.

THE FLASK EXPERIMENT

This experiment was devised by Tancock in his Giggleswick days and is described in his *Starting Astronomy* (8).
(It is also described in (7) and in film strip CGA-192). It
consists of a round-bottomed flask just over half-full of
water and held upside down. It represents the celestial
sphere (see Fig. 1); pupils must imagine that they are inside

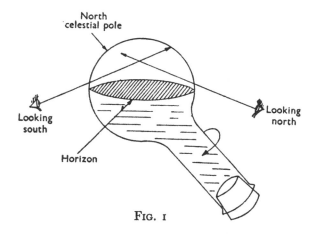

FIG. 1

it, and look through the near side at the far side. Stars can
be added in chalk. A 1- or 2-litre size is suitable for demonstration, but with a keen class it is better to issue a 250 ml.
to each, or each pair. The apparatus is extremely useful
and can be used to show: (*a*) apparent rotation of the sky,
(*b*) rising and setting, (*c*) circumpolar stars, (*d*) large and
small arcs in the sky, (*e*) effect of changing latitude, (*f*)
length of the day in different seasons, (*g*) altitude of the Sun
at noon in different seasons and latitudes, (*h*) angle of rising
and setting, (*i*) morning and evening stars.

PLANE OF THE ECLIPTIC

If any difficulty arises try the old dodge of floating a ball in water, with a pin through the ball to represent the axis (7).

MOTION OF THE MOON

Another opportunity for demonstrations with the ball and lamp, and here the room really ought to be dark. One or two of the children should be brought out to represent the Earth and move the Moon around their heads (7). Failing the demonstration the facts can be taught by drawings, and it is worth while doing them accurately (5) as to some extent drawing must take the place of the usual laboratory periods. The reason for the infrequency of eclipses really must be demonstrated as it involves two different planes, though even intersecting planes can be diagrammed with pieces of cardboard—making three dimensional diagrams with cardboard is mentioned in Burns (5). When studying the Moon's surface put a photograph in front of the class, point out one or two craters, and ask the pupils to model them in plasticine. It is interesting to see how they interpret the effects of perspective and illumination.

THE PLANETS

Do not fail to give the history of planetary discovery, from Ptolemy onwards; it is a fine example of the development of scientific thought. When it comes to paper-work the orbits of the planets can be drawn to scale; circles will be good enough, but two diagrams should be made as Pluto and Mercury will not fit on the same scale on any ordinary paper. If the Elements of the Planetary Orbits (given in the B.A.A. *Handbook* (45)) are available the positions of the planets can be inserted. An extension of this (5) is to deduce in what constellation each of them lies. and hence whether they should be looked for on the evening in question. The class can also calculate the sizes and distances on some scale such as football=Sun, and then make models of them in plasticine or other modelling material. The last lot that I had were made as an exercise in using vernier calipers. The phases of the two inner planets should be shown with the ball and lamp.

CONSTELLATIONS

When dealing with this topic something should be said about their origin and classical meaning. With juniors, do not fail to show them the pictorial representations, and tell them the legends. Apart from just sketching them, accurate maps can be made by plotting the co-ordinates of the chief stars on squared paper (8).

SUNDIAL

Making a sundial never fails to interest, but it needs one of the better classes if they are to understand what they are doing. A straightforward geometrical method of drawing it will be found in Barlow and Bryan (11).

In planning the course for the term, keep an eye on the astronomical phenomena, so that as far as possible the outdoor work outlined in the next chapter can be correlated with the classroom part of it. Give your lesson on the Moon at the time of new Moon, and if you can talk about planets when there is one about, so much the better. Similarly the constellation study should not be done all at once, but bit by bit as they come into view in the evening sky.

SIXTH FORM

A short syllabus has been given and should hardly need comment; teaching astronomy at this level is much the same as teaching physics. In order that the course may serve its purpose of widening horizons, it is essential that it be supplemented by reading, and a suggested list of books appears in the bibliography. Your readers should be put on their guard against the apparent finality with which some authors write. In cosmogony, for example, much of the work of Jeans, so persuasively put over in the 1920's (24, 34), no longer stands as new knowledge has made it untenable. To-day Hoyle writes persuasively (35), but in this case it is debatable whether his ideas have yet sufficient support to convert them from 'possible' to 'probable'. The current values for, say, the diameter and the age of the Earth differ very widely in probability; try to prevent your pupils from giving them equal weight. Ideas on teaching the sixth form were given by Westaway in his *Science Teaching* (1), and although the content of his proposed lectures will need bringing up to date, the theme and general treatment are still sound. Resulting from their reading the pupils should write longish essays, or give lectures to their fellows in class or in society meetings. Another idea that I have seen used is to give one of them, a few days ahead, the lesson notes of a film strip, and get him to show and explain it to the rest of the form. Here are a few suggestions for essays or talks to be prepared by pupils from their general reading and simple research:

1. The astronomy of the ancient world.
2. Copernicus and his successors.
3. Astronomical developments of the eighteenth century.
4. The growth of the astronomical telescope.
5. The spectroscope in astronomy.
6. Man's horizon is ever receding.
7. The astronomy of the Bible.

8. Shakespearean astronomy.
9. Classical legends in the heavens.
10. The poets; astronomy or astrology?
11. Great names on the Moon.
12. Applications of radio to astronomy.

Preparing the historical essays should not be limited to an astronomical book list, but should include such works as Westaway's *Endless Quest* or Dampier's *Shorter History of Science*; the astronomy must not be treated in isolation, but fitted into its background. No. 7 above is the title of an old book by E. W. Maunder; it might be dug out of a library somewhere, though possibly the essay should be prepared first. Nos. 8 and 10 could be set as questions; give some references (the literary staff can supply plenty) and ask for criticisms of the science (or otherwise) therein. For the teacher to check No. 11 there is a publication called 'Who's Who on the Moon',* but it would make the work too easy to let the pupil get hold of it. At the time of writing the first edition there was only one textbook on No. 12, and that a difficult one. Now there are a number such as (39) and (39*a*). The Royal Astronomical Society has published a useful summary up to 1954.† See also references (52–54).

Some of the essay subjects are suitable for the non-science specialists. The general aims of science for non-scientists were given by McKenzie some years ago,‡ and a sub-committee of the Science Masters Association drew up a syllabus.§ This syllabus was afterwards incorporated in the *General Science Report* (4) and can also be found in (3). The

---

* *Mem. Brit. Astr. Ass.* Vol. xxxiv, Part I (1938).
† *Occ. Notes R. Astr. Soc.* No. 16 (1954).
‡ 'Science for Arts Specialists', *Sch. Sci. Rev.* No. 95 (1943).
§ Report of Sub-Committee, *Sch. Sci. Rev.* No. 99 (1945).

content of the astronomical part is included in our basic syllabus and extension 1, but the emphasis is different. The historical side is more important, and McKenzie includes the actual words 'the abandonment of the sun-centred Universe' as an important step in the main theme of the development of scientific thought. Some mention should also be made of astrology, its influence on life and literature in various periods, and the stimulation and or hindrance that it made towards the genuine science of the heavenly bodies.

PRONUNCIATION

Teachers of many subjects run into the difficulty of choosing between common usage and etymological correctness, and there is no short answer to it. So much depends on the standard of general scholarship in the class concerned and teachers must use their own judgement in the matter. To quote two astronomical examples: Uranus should be pronounced with the accent on the first syllable, but the long a as in Uranium is much more common, especially now that Uranium is so often mentioned in the newspapers. The astronomer speaks of the star Sirius with the i as in siren; there are many people to whom this name would mean nothing but who are quite familiar with 'Sirrius'. Astronomical custom is an added difficulty. The Latin ending -ae is used by the astronomer like the -ee in tree, whereas for the last forty years of school Latin it has been like the -ye in dye. Thus the plural of nebula is nebul*ee* in the astronomical meeting and nebul*ye* in the classroom. The same occurs with the -i ending, -ye to the astronomer and -ee in the classroom. Personally, in my school work I always agree with my Latin colleague and generally with common usage also; the few pupils who will go on with astronomy seriously can have these problems pointed out to them in the sixth form. Pronunciations of many star and constellation names are given in (17, 40) and the commoner ones in (12).

Experienced teachers should omit this paragraph, it is for beginners only. You will probably find that you cannot lift the various suggestions in this chapter and the next direct to your own classroom. Not all teachers have a flair for breaking away into a new subject for which they have not been specifically trained; not all pupils are industrious, soaking up knowledge as blotting paper soaks up ink; not all schools have the same kind of discipline; not all classes, even in the same school and with the same teacher, have the same standard of application and behaviour; in some over-crowded classrooms it is not possible to bring children out to play at Earth and Sun; some youngsters can find alternative occupations when the lights are out. This is not an infallible cookery book; the most that the writer can do is to give ideas that he and others have found useful. You must adapt the ideas to the peculiar needs of yourself, your pupils and your school, and so make them your own; until they are your own you cannot put them over successfully. One last word: do not worry if you are not very well up in astronomy yourself. It is possible for teacher and pupils to work up a new subject together, provided that the pupils know—they will enjoy helping you!

CHAPTER III

IN THE OPEN AIR

Astronomy is not an armchair (or school-desk) subject; the learner simply must make some personal acquaintance with the objects of his study. Herein lies the greatest difficulty of astronomy teaching. Not only is it impossible to bring the exhibits into the classroom, but neither the day nor the hour when they *may* be observable are under the teacher's control, and all too often the weather defeats any arrangements that have been made. There is some outdoor work that can be done in daytime, though this may demand some elasticity in the time table. It is sometimes possible for children to come back to school after dark, though the ever-increasing centralization, with the inevitably enlarging 'catchment area' and the school bus, is making that difficult. The situation is easier in boarding-schools, though even there I find it difficult to do much beyond the limits of my own boarding-house except when the days are shortest and darkness comes at 5 p.m.

It is clear, then, that something must be devised for the pupils to do independently in their own homes or boarding-houses, and some suggestions of this nature will be found in the following pages. Some of them they can all do and write up individually in their own note books. Others may involve some simple equipment that is not available all round. I have used exercises of this kind; two boys were in charge of each piece of work, and after some period of time, say three weeks, each pair reported to the class what they had achieved. One pair would be able to exhibit a series of drawings showing the changing configurations of Jupiter's satellites; another described the sunspots he had seen, and so on. Some teachers, where there is a school telescope, make themselves available on certain evenings for any of their pupils who choose to come. All those with whom I have had an opportunity to discuss this have found, as I have, a ready response, even in day schools. The response in one day grammar school was almost embarrassing—the wretched science master could hardly ever get home!

The remainder of this chapter will be devoted to exercises of various types. Most of them appear in the books already quoted: a course of practical work is given in (2); a chapter on this aspect in (8); exercises at the ends of the chapters in (6) and (7); (12), though not a textbook, was specially written to help the youngster to help himself.

GROUP I
(Individual work not requiring the presence of the teacher or the use of a telescope)

CONSTELLATIONS

The pupils can look for these at home. In school they can copy them down from the blackboard, and at this stage it is important to make clear how large they will be in the sky. A fully extended hand at arm's length is just about 20°, and a line 20° long should be added to the drawings and labelled 'hand-span'. A finger is about 1° and could be used for small groups like the Pleiades. In some cases it will be possible to tell the class to make some addition to the drawing while they are looking at the sky. Get them to record in some way the direction in which they saw the constellation—'over the next-door chimney when I was standing by the dustbin' may not mean much to the teacher but it does to the child. If the observation was repeated '. . .

when I got home but wasn't there at bedtime' an important lesson would have been learnt. Amplify this lesson by having the direction written down, each child doing it in his own way, and done again after a few weeks. As progress is made, maps of the sky could be made available; a simplified version of *The Times* map reproduced on the duplicator would do, but it would be a help if a set of *Stars at a Glance* (16) were available.ᶜ In any case the class must be taught in the classroom how to use it, and the teacher must be prepared for some pupils who will never be able to. Some people, old and young, never seem to see the constellation patterns, and there are others who cannot to begin with but have no difficulty after they have once been pointed out by someone else. Some people, old and young, have no sense of direction either and are incredibly stupid if you ask them which way the street runs; if some of your pupils do not know which is the south side of their homes ask them to see where the Sun is during the Sunday dinner. Once the Pole Star has been located this difficulty is removed, and the Plough and Pole Star are likely to be among the first things to be taught. Get your pupils to note regularly the position of the Plough relative to the Pole, using the clock notation for simplicity; try to link up these results with the stars 'over the next-door chimney'. Some of the class might like to make a nocturnal (5), an apparatus for telling the time by the Plough. The same book gives geometrical problems about the positions of the constellations, and this classroom work lends itself well to outdoor checking.

THE MOON

The obvious thing to do here is to watch for it at the beginning of the month, and record the direction, time and phase regularly throughout the month. Get them to find it in daylight too, and link up all this with the corresponding classroom work. A useful exercise is to prepare with the duplicator a chart of the constellations through which the Moon will pass. Issue these to all pupils and get them to insert the position of the Moon as often as possible for the term. Put the question 'find out how many days it takes the Moon to get back to the same place among the stars?' Results with these charts will be disappointing at first, but good results can be got with perseverance.

THE PLANETS

The same charts can be used for following the motion of the planets, and this exercise seems easier than the Moon one, possibly because the moonlight makes the stars difficult to see. If there is a planet in a prominent constellation, set the class to find the constellation without mentioning planets; many of them will fail to recognize the constellation at all, but there is sure to be someone who will tell you that your map was wrong! Watch for evening and morning stars, and if the opportunity occurs arrange a watch for Mercury on the appropriate dates; evening apparitions in the early spring are the most favourable. For planetary data see *Whitaker's Almanack* or (45).

OTHER USEFUL EXERCISES

Set someone to watch the variable star Algol; choose a patient person, and do not start him until somewhere near the date of a minimum (*Whitaker* or (45)). I have known a boy work for a whole month and put seventeen 'normals' in his book before seeing a change, but this is exceptional and most boys would have packed up long before that. The variables β Lyrae and δ Cephei can also be studied with the naked eye, but these are not suitable for juniors other than the very occasional gifted enthusiast. Warn the class about meteors and ask them to be sure to report to you any that they see. If there is a sundial available a regular comparison of its reading with correct mean time is a good exercise provided that it is kept going for a long enough period. Another daylight job is for a child who has a pocket compass to note the direction and time of sunset (and sunrise in the winter) once a week for the school year. You will have to jog his memory from time to time! Perhaps it would be easier to provide the compass and change the observer once a month.

GROUP II
(No telescope, but help probably required)
FINDING THE MERIDIAN

This is done, of course, by studying the shadow of a vertical post on a level playground. An alternative which I have used and prefer (having at the time a south-facing laboratory) was a retort stand on a laboratory bench at a time of year when the Sun was low enough to shine well into the room. The meridian can be found by marking the shadow at the calculated local solar noon, but for teaching purposes bisecting the angle between two equal shadows, observed about an hour each side of noon, is a sounder method.

LONGITUDE

When the meridian has been found it can be used to time the transit or meridian passage of a heavenly body, using two thin posts or sighting slots. The time of transit for Greenwich will be required, and the difficulty is to explain to juniors how that was obtained. Using the Sun will involve a correction for the equation of time, but if the class have already done the sundial-clock comparison the equation is just the sundial error of which they are already aware. *Whitaker's Almanack* gives the equation of time in the sense apparent minus mean, so that the transit at Greenwich occurs at mean noon plus the positive or negative quantity printed. The *B.A.A. Handbook* (45) gives the Right Ascension of the true Sun, but not the equation of time. To use this or the R.A. of a star or planet, which is the sidereal time of the transit, it must be converted into mean time by use of tables that are available in both reference books.

LATITUDE

This can be found from the altitude of the Pole Star. A clinometer can be improvised with a strip of wood about a foot long having sighting slots fixed perpendicular to it at each end, and a protractor and plumb line in the middle. While one pupil views the star a second uses a dim torch to read the protractor against the plumb line. The Pole Star is not quite at the celestial pole, but the error is not likely to exceed the experimental errors of two twelve-year-olds with an improvised clinometer. A table of corrections will be found in *Whitaker*. Do not expect high accuracy with this set of exercises. Point out to the class that with the apparatus available they cannot be more accurate than the nearest degree, whereas the navigator on a ship is a trained man using an instrument reading in minutes of arc. Show a picture of a sextant if you can.

ALTITUDE OF THE SUN

This can be found with the same clinometer provided that a suitable dark shade is available; *ordinary sun spectacles will not do*. These results can then be correlated with the sunset readings mentioned in the last section.

GROUP III
(*With small telescopes*)

This work cannot be classed as either individual or demonstration, it will probably include both. Any optical aid is better than none, and the educational value of 'making do' with small or improvised apparatus cannot be overstressed. Some pupils will have something of the kind at home, or in their possession at school, and they can be set individual tasks to do and report on. Where the school possesses something bigger and better it should not be used early in the course, and in fact the school should possess one or more small instruments as well. By small instruments I mean binoculars, old-fashioned field glasses, and telescopes in the 1- to 2-inch range. It is important that the youngsters should learn to appreciate how much can be done with the small before they are introduced to the large. If they are allowed to use the 4 in. right away they tend to despise anything less and will give up astronomy as soon as it is no longer available to them. Two things must be told them at the outset; they must stay in the dark for a little while before they can expect to see well, and the instrument must be steady. Small telescopes should have a stand of some kind; a laboratory retort stand will do, and stands can be made with Meccano—my first one was. When using binoculars the hands must be steadied in some way, such as resting the elbows on top of a wall or sitting in a deck chair with the hands resting on a broomstick held between the feet. Here, then, are a few of the things to do, individually or collectively, with limited optical equipment.

THE MOON

This is the most satisfactory object to study and a never-ending source of interest. Study should begin with the thin evening crescent, and the young observers should try to draw any markings they can see, particularly in the regions of the terminator where shadows emphasise the relief. Each night the terminator moves further east and new objects come into view and can be added to the drawing. Even binoculars show far more than can be drawn in school, so some guidance may be needed to direct attention to objects of interest. Simple maps of the Moon will be found in many of the books mentioned, including (16), enabling some of the objects seen to be given their names. One reference book (15) gives a map for each day. With the larger instrument it is more profitable to concentrate the attention on one particular formation or region and note how its appearance changes from night to night as the illumination changes.

THE SUN

The Sun can be observed directly, preferably when low in the sky, by using a suitable dark shade on the telescope eyepiece. Improvised shades are always rather a danger and it is much safer to use only the projection method in school. Fig. 2 should make the arrangement clear; orient the telescope so that it casts no shadow on the first screen, and then focus a sharp solar image on the second. The position of any sunspots seen should be recorded by diagrams in a notebook, and the observation should be made every fine day for at least two months. For about two years around the minimum of the eleven-year sunspot cycle this could well be left out of the junior course—the observers will get tired of having nothing to report. Seniors, on the other hand, could very well keep a watch for the first high-latitude spots indicating the commencement of the new cycle. The last minimum was in 1954.

VENUS

The phases of this planet can be followed with a 2-inch telescope, nothing smaller, and it is important to observe while there is as much daylight as possible. Venus is very brilliant and in a dark sky it is difficult to get a really sharp image with ordinary instruments. Shortly after the evening elongation and before the morning one the planet presents a crescent and has a large apparent diameter; at this time the phase is unmistakable with binoculars, or almost any instrument.

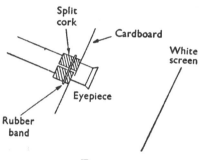

FIG. 2

JUPITER

The four large satellites are only just beyond naked-eye visibility, and any optical aid will show them in a dark sky. Here it is necessary to emphasise the point about keeping the instrument steady. A useful routine exercise that my pupils always like doing is keeping a daily (weather permitting!) record of the configuration of these satellites. It is interesting, too, to see one vanish into eclipse or reappear, but this can hardly be called a class exercise as it is an instantaneous event and only the privileged can see it. The times of these eclipses are given in *Whitaker* and (45) gives the daily configurations as well. Good small instruments will show the existence of surface markings, but do not stress this at the junior level as some could see it and others not—the class as a whole would not be convinced.

OTHER OBJECTS

This is really a list of recommended demonstrations, though most of them could be done by the more able pupils if they were told how to find them.

Compare the appearance in the telescope with the naked eye view, first for a planet and then for a star of about equal brightness; draw attention to the disk-like effect of the former even with very small instruments.

Show the rings of Saturn; a 2-in. telescope will do it when they are fully open. If a better telescope is available have a drawing made each year and display the series in class sometime.

My stock double stars are β Cygni and Mizar, which are quite convincing with anything from 1½-in. up. For smaller instruments I use ν Draconis. Before showing the star explain what position angle is, and ask the observers to describe what they see in terms of a clock. This is a useful check that they are seeing the double star. Ask for their opinions about colours, I once tried out a group of boys on a number of stars* and found that their estimates differed a good deal from the textbook colours.

The Pleiades; have them counted with and without the telescope.

The double cluster in Perseus; a sweep of this area between α Persei and Cassiopeia should also demonstrate the starry nature of the Milky Way. Praesepe.

* Estimates of Star Colours by Inexperienced Observers', *J. Brit. Astr. Ass.*, Vol. LVIII, p. 155, 1948.

Nebulae in Orion and Andromeda.

After showing them, but certainly not before, draw attention to the observatory photographs and point out the essential difference between them. Most youngsters are interested to hear that the Andromeda nebula is the most distant thing that the naked eye can see.

These suggestions are for the ordinary middle school classes; work for seniors and better telescopes will be mentioned at the end of this chapter.

GROUP IV
(*With cameras*)

It is unfortunate that photographic material is more expensive than it used to be, for it discourages the school boy from using it for school work. However, if any camera users should propose to do so here are a few simple experiments worth doing. Generally speaking this work cannot be carried out by town dwellers as real darkness is essential.

ROTATION OF THE EARTH

Prop the camera facing the Moon and leave the shutter open for about an hour. A trail will be obtained, and by comparing the width with the length the pupil can estimate how long it takes the Moon to cross its own diameter. Similar trails can be obtained for fairly bright stars, and show clearly that the paths of stars across the sky in the South are parts of concentric circles. If the camera be aimed at the pole on a favourable night and a really long exposure made, say 4–6 hr., the centre of the circles can be located and be seen not to coincide with the Pole Star itself. Avoid all stray lights, use the largest stop and give a full development. For the Moon trails f/11 is enough.

MOTION OF THE PLANETS

If there is a planet in the field of the star trails its motion can be revealed by superimposing two negatives taken at a suitable interval apart. Pupils are quite impressed at being able to detect the motion of Mars in as short an interval as a couple of days.

THE MOON

With a 3-inch telescope quite pretty little photographs can be made at the focus without a camera. Some means

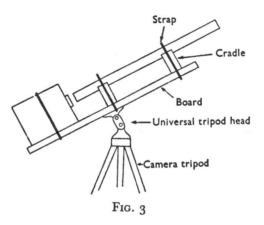

FIG. 3

must be devised for attaching a plateholder to the eye end. At one time I had a small screen with the appropriate slides on it, and this was attached to the suncap thread of an eyepiece collar from which the diaphragm had been sawn out. However, teachers can design something suitable for the instrument in question. Focus on a piece of ground glass; hold a piece of cardboard in front of (not touching) the objective; put on the plateholder and draw the slide;

wait for a moment for vibration to cease, and make the exposure by moving the cardboard for one second. The resulting image is small, but can be sharp enough to make a good miniature lantern slide.

Trying to combine telescope and camera always interests, but the method of doing it must be left very largely to the school concerned as it depends on the nature of the apparatus used. Fig. 3 shows just one arrangement emphasizing the need for a rigid connexion between the two parts. This time the eyepiece is left in place and the camera lens can be—in fact I always do this. If it is a plate camera with a focusing screen, use the telescope rack-work to do the focusing. Without a screen the camera must be set for infinity and the telescope in normal adjustment, i.e. set for parallel light; this ought to give a sharp picture, but if it does not a correct setting can usually be found by trial and error. Use fast plates or film (I like Ilford HP3), an exposure not exceeding 1 sec., and a cable (not trigger) release for making the exposure. For taking the Moon without the camera as outlined earlier very fast plates are not so necessary.

THE SUN

This can be photographed in the same way as the Moon, but use a slow plate, the fastest shutter speed, and stop down the objective by covering it with a card having a central hole not more than an inch in diameter.

More elaborate photography can be done in schools, but it is more in the nature of sixth-form or society work. An account of an astronomical camera will be found in the *Science Master's Book*,* and the practical aspect of

* Series I, *Physics*, p. 68; also reference (48).

amateur astronomical photography was the subject of a recent Presidential Address before the British Astronomical Association.*

Special occasions should not be wasted. If a comet comes within reach of the naked eye it should be followed night by night, and it might be possible to duplicate and issue simple star maps of the area. The comet of 1942, moving rapidly across the Plough, was an ideal object, but such convenient visitations are rare. The small telescopes are quite suitable for studying bright comets. Even rarer in southern England is the aurora, but do not fail to ask 'who saw it?' if it does occur. Scottish schools, on the other hand, could be set to look out for it.

Eclipses form important special occasions. For solar ones see that the children's eyes are properly protected. Sufficiently dark filters (much darker than sun spectacles) can sometimes be obtained, but I have found that the ordinary laboratory colour filters of 'theatrical' gelatine can be used if primary colours such as red and blue are superimposed. Smoked glass is satisfactory as a protection, but it is surprising how soon it is noses, hands and textbooks that are smoked. On the occasion of the solar eclipse of 30 June 1954 school observations† included the following: timing and drawing the progress of the eclipse; observing irregularities on the Moon's limb; variation in temperature; variation in intensity of daylight; behaviour of birds; reaction of flowers; colour of the sky; search for planets and bright stars. There is plenty of scope for working on a good eclipse. At a lunar eclipse the progress can be drawn and colour should be noted. A photographic record

* By E. H. Collinson in *J. Brit. Astr. Ass.*, Vol. LXV, No. 1 (1955).
† *J. Brit. Astr. Ass.* Vol. LXV, No. 1 (1955), gives a summary of the observations made by eight schools.

can be made in the same way as for the Moon trail, only instead of leaving the shutter open all the time take a series of separate exposures of 1-3 sec. On development there is a row of images showing the progress of the eclipse. If a telescope is available see if your pupils can identify any formations in the eclipsed region.

With senior boys and better telescopes the field of observation is very much wider. If a suitable timepiece is available the observation of lunar occultations can be undertaken. Predictions are given in *Whitaker* and in (45); detailed instructions for doing it have been published in the B.A.A. *Handbook* for 1949. Many more double stars become available and many lists have been published, as in (15), (17) and (18); there is a short selected list in (8). Here a word of warning; do not expect a beginner to work at the limits of performance, but allow a wide margin so that he will not be discouraged by failure. A 1½-inch telescope is supposed to separate a double star with components 3.04″ apart; with a perfect instrument, a perfect night and a skilled observer perhaps it will. With this size of telescope do not set a beginner anything closer than 20″; with a 2 in. call the school limit 10″ and a 3 in. 5″. Similarly avoid large magnitude differences, as the bright star will make the fainter difficult to see even if it is far enough away. Experienced amateurs, if they ever read this, will writhe in horror, but never mind! If the telescope is large enough it may become possible to study planetary detail, but now we are getting beyond the terms of reference of this book—this is not a practical handbook for the capable amateur (see (42)). The teacher who is fortunate enough to have the better telescope and the better boy or girl to use it should read the aims and objects of the B.A.A. (46); this contains advice, written by experts, on what to do and how to do it.

There is, however, one kind of exercise that I have found useful with the keen senior, a simple piece of 'research' where it is important that he should not know in advance what he is expected to see. He is provided with an instruction sheet, with spaces for results. One of these, intended for use with a $3\frac{1}{4}$-in. refractor, is reproduced below. He should find three double stars; if the telescope is good enough he will find that one of them is quadruple and will just glimpse a nebula; if he is honest and puts down what he really sees instead of what he expects to see he will 'discover' a variable star. Lyra is exceptionally good for this purpose; I have also used Cygnus, Andromeda, Perseus and Cepheus, but none of these offer so much as Lyra. No doubt some of my readers will be able to think of others.

NAME.............................. DATE............ TIME............

LYRA (the Lyre)

*

* *

*

*y

*x

Letter the stars in the diagram from A to G in order of brightness. If two are equal use the same letter twice and omit the next one.

Examine each star, and also the space between x and y, first with the naked eye and then with the telescope; note any peculiarities or colour contrasts:

....

....

....

....

....

....

....

On two other evenings check your lettering for brightness:

DATE................. REMARKS ....

DATE................. REMARKS ....

Put here any other comments regarding your observations; refer to weather conditions:

....

....

....

4

THE SCHOOL TELESCOPE

The preceding chapter will have shown that some kind of telescope is needed, and that none is too small to be useful. There are on the market some 'toy' telescopes for children; these may be exceptions to the rule, but I have not tried one. If there is no telescope at all but there is a physics laboratory, a raid on the lens drawer should produce a pair of suitable lenses with which a simple telescope can be made. Other requirements will be a pair of cardboard (or rolled paper) tubes, one to slide over the other, and some means of fixing the lenses—it should not be necessary to go into detail when writing for science teachers. The pair of lenses I have found most satisfactory for this job is a 65 mm. × 50 cm. for the objective and a 25 mm. × 5 cm. eye lens. Simple lenses are used, so they would not cost very much to buy specially, and although not achromatic the telescope will show lunar features, sunspots, the satellites of Jupiter and certain double stars. A difficulty is that the exit pupil is about two inches outside and, in spite of telling, the boys will bring the lens close to the eye. I now fit an extension paper tube to keep the eye back to the proper distance. An ordinary retort stand can be used for supporting the telescope as it is very light, or it can be mounted on a post or tripod if desired.[d]

In the 1-inch to 2-inch class all sorts of telescopes exist, folding or otherwise, and some of these are provided with a light tripod too low for the astronomer, or with a table stand ideal for the lounge window of a seaside boarding-house. Some use can be made of the former as a basis for a new tripod. The latter frequently incorporates a vertical

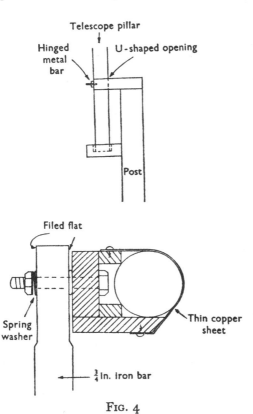

FIG. 4

pillar which can be utilized for mounting on top of a post placed permanently in the garden or playground. Fig. 4 should give all the information needed. In deciding the height of the post remember that your pupils are not as tall as you are. Hold the telescope at 45° to the eye of an average user, and get someone else to drop a tape from its centre of gravity to the ground. If the telescope has no stand at all it is not difficult to devise and make a cradle by by which it can be mounted on the post. The one illustrated was made recently to carry a 2-inch spotting telescope; the dimension perpendicular to the paper is about 4 in. and other measurements depend on the diameter of the tube at its centre of gravity. It is important that the axis of the bolt should pass through the centre of gravity of the telescope, so that, with the spring washer under the nut, it will stay put at any altitude. A final word about these terrestrial telescopes: it is not worth while trying to adapt them to take astronomical eyepieces, for they will perform better with their own erecting eyepieces.

Let us suppose that a telescope is to be acquired for the school. Then what sort and how big? It is hardly necessary to argue here about the relative merits of the reflector or refractor. In professional spheres each has its own particular uses, but in school either can do the job and convenience becomes important. The refractor is simple, easy to understand, requires little maintenance, and points in the direction that the pupil is looking—he or she can soon learn to use one individually. The school telescope, then, should be a refractor, though reflectors will receive further mention later. Now for the second question, how large? For instructional purposes in school I have never felt the need for anything larger than a 3 inch, and quite a lot can be done with it.* It is easy to use and has the great advantage of being readily portable; it can be taken out into some open

* E. A. Beet, 'Possibilities with a Three-inch Telescope', *J. Brit. Astr. Ass.* Vol. LVIII, No. 4 (1946).

space where the pupil is under the whole sky and ın find his bearings as well as obediently carry out the order to 'look through this hole'. A 4-inch is nominally portable too, but moving it is a heavy job, and anything larger than that must be permanent and introduces the observatory problem. An equatorial mounting is very nice, but it costs a good deal more and in a portable instrument much of the advantage is lost. The ordinary horizontal and vertical movements of the altazimuth will do all that is essential, and its use is good training for the beginner. Such a telescope will cost from £35 new, with an 'educational' class of objective, and would be over £100 in the highest optical quality. Thus telescope buying should not be rushed into light-heartedly. Read Captain Ainslie's chapter in *The Splendour of the Heavens* (40) describing the amateur and his equipment. If you have access to the B.A.A. *Journal* read Mr Holborn's address on 'The Beginner's Telescope'.*

A telescope bought new from a reputable maker is probably optically sound, and may be guaranteed. There are a number of good firms who deal in second-hand instruments and will give their customers help and advice, but with any second-hand goods, particularly that privately advertised or appearing in sale catalogues, the purchaser must take care. The processes of testing and adjusting a refractor are described quite fully in both the references just quoted and also in (41), and in any case would be out of place here. As a preliminary guide, focus the telescope on a star of moderate brightness well clear of the horizon and make sure that the image is free from flare, i.e. without a little tail like a miniature comet. It should be a small circular disk with one or two complete diffraction rings, and on throwing the telescope out of focus a little the disk

* *J. Brit. Astr. Ass.*, Vol. LVIII, No. 1, pp. 6 ff.; also reference (47).

should enlarge in exactly the same way whether racking in or out. If the telescope does not satisfy these tests it may be all right, but needing only a little adjustment, but on the other hand it may be a poor instrument, so take advice before completing the deal.

Now for the stand, the desirable features of which are illustrated in Fig. 5. The table stand has already been condemned; the garden tripod is essential unless you propose to make one. Some tripods, however, are crowned with the pillar and compass joint like the boarding house telescope previously mentioned. One defect is that this mounting is really meant for spotting telescopes which are used horizontally, so that the perpendicular from the centre of gravity is not very far from the axis of rotation. At high inclinations this is no longer the case and the telescope will not stay put—tightening the joint is a poor solution! Steadying rods deal with this trouble, but not with the other, which is that the zenith cannot be reached. The illustration shows a telescope pivoted through its centre of gravity, and on bearings that are off-set from the pillar so that, except when the telescope is over a tripod leg, the zenith can be reached. This kind of telescope will stay put without steadying rods, but with a school instrument it is desirable to have them as some children will bump their heads on the eyepiece instead of looking through it. Even steadying rods will not protect you from the loss of your setting, as it will not be long before somebody falls over a tripod leg! The only answer to this is an orderly queue, and when each observer reaches the telescope try to stop him or her from grasping the tube with both hands like a 'pop' bottle and starting up vibrations. Train them in daylight to put an eye to the telescope without touching anything. Organizing an excited class of small boys (and

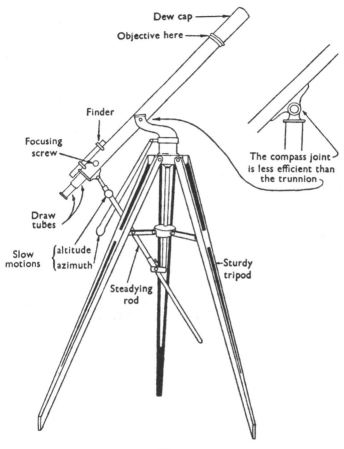

Dew cap

Objective here

Finder

Focusing screw

Draw tubes

Slow motions {altitude azimuth}

Steadying rod

Sturdy tripod

The compass joint is less efficient than the trunnion

FIG. 5

presumably girls too?) in the dark requires more than patience—but it is worth the effort every time.

But to return to the telescope. The slow motions are not essential, but are very useful, for with a little practice the teacher gets to know how fast to move them to keep an object in the field while quite a succession of pupils take their turn. The finder is also useful but not essential; if you cannot get your eye to it while a pupil is at the main eyepiece try using a 90-45-45 prism from the physics laboratory. The dew cap is essential and should be at least four or five diameters in length. It can be made of a roll of paper painted matt black inside. If you have a metal one it must be lined with blackened blotting paper. The finder should have a dewcap as well. Put the dewcaps on before bringing the telescope out of doors; if you get the lenses dewed you have finished for the evening, for a good objective must not be wiped as if it were a misty window pane. Before taking indoors again, remove the dew caps and very quickly replace them by the permanent dust caps. Keep dust caps clean, so that cleaning the lens is a very rare occurrence. If it must be cleaned set the telescope horizontally and dust the lens with a soft sable painting brush. Then wash with cotton wool, a trace of soap, and a minimum of water. Dry and polish with an old, soft and very clean cotton handkerchief.

Two or three eyepieces will be required, and are probably supplied with the instrument. Do not be tempted by high powers; for inexperienced observers and average weather a power of about 30 per inch of the object glass, say 100 for a 3-inch telescope, is as much as is really worth while. The lowest power should be as low as you can get, and must at least be capable of showing the whole of the Moon's disk with something to spare.

There are advantages in having an equatorial head: following the object can be done with one control only, and this can be mechanically driven: if fitted with circles objects such as the planets can be located conveniently in daytime; long exposure photography becomes possible. On the other hand it is much more expensive, a 3-inch hand-driven type costing as much as £200. If it is going to be portable there is the added weight and the greater liability to damage, and to some extent the value of the circles will be lost. A portable equatorial should always be used in the same place, and the position of the tripod feet marked in some way on the ground. The method of setting the instrument in the first place is given in (41), where the whole question of mountings is discussed, including suggestions for converting an altazimuth into a rough equatorial (this is also given in (17)). It is possible for a simple equatorial stand to be made in the school workshop, and this has been done (50). Making the telescope as well is not usually regarded as being an amateur job, but that has been done too.* If there is to be a school observatory on a permanent site the full value of the equatorial can be realized, and with telescopes of 4 inch and upwards it is 'spoiling the ship for a ha'porth of tar' not to have it—though in this case the 'tar' is rather more expensive.

If the school already has a small refractor for instructional work, and now want a second and larger instrument for the senior astronomical society, then perhaps a 6- or 8-inch reflector might be considered. It will need more looking after; it is less convenient to use; the eyepiece is liable to get into inconvenient positions. As it is not to be used for mass demonstrations these troubles are perhaps less serious

* A. E. Watson, 'A School Astronomical Telescope', *Science Masters' Book, Series II, Physics*, and reference (49).

than they sound, and there are advantages. For their size reflectors are much less expensive than refractors, a new 6 inch costing from £70; they are considerably shorter and therefore easier to house; they can be successfully made by amateurs. An instrument produced in school has been described in *Sch. Sci. Rev.* (51), though in this case a finished mirror was available. With a sufficiently keen (and lei-sured!) science or handicraft master and some industrious and persevering boys it should be possible to do the whole job in the school workshop. *Constructing an Astronomical Telescope* (43) gives a general idea of what is involved and how to set about it, but if it is decided to go ahead, then it is essential that the standard work on this subject by Ingalls (44) be consulted.[e]

Of auxiliary apparatus the camera has already been men-tioned. If the telescope is an equatorial a camera with a lens of wide aperture can be attached to the tube level with the centre of gravity and used for longish exposures of star fields. If the camera is attached to the eyepiece a counterpoise must be put on the other end or the unbalanced telescope will put a strain on the driving mechanism.

A star diagonal is very useful when showing objects at a high altitude. A single observer can get to the ordinary eyepiece, possibly with some discomfort, satisfactorily if he is not in a hurry. With a queue of pupils, however, the view-point must be conveniently placed and the diagonal does it. Remember that it introduces lateral inversion.

The small direct-vision spectroscope to be found in many physics laboratories can be adapted as a solar spectroscope. The telescope is used without an eyepiece and arrangement must be made to attach the spectroscope in such a way that the slit can be seen and the solar disk sharply focused upon it (see Fig. 6). It is assumed that the principal focus of the objective is well clear of the telescope body, i.e. considerable drawtube is normally used. To look for prominences the slit must be placed tangential to the limb, and held there while it is opened. It can be done with a 3-inch altazimuth and your sixth form might like to try it, but manipulating the telescope is difficult, and the Sun's image is very small. This is really work for a larger and equatorial telescope.

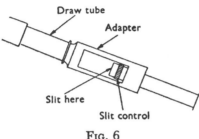

Fig. 6

The slitless star spectroscopes of the McLean or Zollner variety can also be used with a 3-inch telescope, but they too need something larger. It is just about possible to distinguish between A, G and M stars, but there is insufficient light to do much more, and again following is difficult. Spectroscopy is really work for the school observatory.

Mention of the school observatory brings up a rather thorny question with which to complete this chapter. An observatory is not really necessary but it is very nice to have it. The ordinary school instruction can be done, and I think better done, out in the open, but with an observatory and a good permanent telescope the senior pupils can attempt more ambitious things and go farther along the road to becoming able amateur observers.

There are now quite a number of observatories in schools, varying widely in scope and equipment from a 15-inch equatorial downward. This big one is at Stonyhurst, where in 1948 the school took over what had formerly been a separate organization. The Temple Observatory at Rugby has already been mentioned, and I have seen some excellent photographs made there. The Rugby observatory is also equipped for making double star measurements, and for the use of the spectroscopes just discussed. In the various school observatories that I know of the telescopes are in the 4–8 inch range, usually refractors, though I also know of at least one 9-inch reflector. Some of the observatories have been built by members of the school with little outside help, and this in itself has an educational value. One of the most recent in this category is that of Kingston High School, Hull,* opened in 1953 by the Astronomer Royal. This is a brick building with the familiar dome, and adjacent there is an ante-room where preparations can be made—lighted, heated and even equipped with a telephone!

The uses of the school observatory vary. Some are mainly for casual visitors who come along on certain evenings if fine, and certainly some of the observing time should go to this end. I heard it said once, I have forgotten by whom, that everyone should see Saturn's rings at least once in a lifetime, and I do remember one lad of 18 exclaiming as he looked at this planet 'well I never knew it was real—I thought it was just a decoration on the front of *Old Moore*!' A part of the time should go to the instruction of the junior astronomy class. The remainder should go to the science sixth or astronomical society, and is probably the main use of most of them. Thus it appears that an observatory is a desirable thing; it is also another thing that must not

* *J. Brit. Astr. Ass.* Vol. LXIV, No. 6 (1954).

be embarked upon too lightly, for there are difficulties to be honestly faced.

Suppose that the school has been offered the gift of a 6-inch refractor, shall we accept it? It will have to be housed, in quite a big way too. Will the school provide the funds to build an observatory, or have you the time, talent and materials to build it without outside help? Perhaps the building is included in the gift; what about it? Who is going to look after it? A telescope is a valuable and delicate instrument that cannot look after itself. The observatory and telescope must be kept scrupulously clean, the working parts must be kept oiled and in running order. The detachable fittings and auxiliary equipment must be suitably stored and cared for. Even the school holidays are long. Well, Mr Snooks has agreed to care for it, so that is settled. But is it? It may be a case of basing policy on present personalities, and he is not immortal. Are the governors prepared to make the care of the telescope a condition of employment for his successor, or is the passing of Snooks going to mean years of rust, decay and pilfering? It can happen, and it does happen. I have heard from men who have had the task of resuscitating a school observatory when the school has suddenly discovered that they have got one somewhere. Observatories, like other buildings, and telescopes, like other machinery, need maintenance; are the governors prepared to provide funds to maintain them? As I write I can think of one observatory that is defunct because the building is no longer weather-proof and the money for repairs is not forthcoming. How much will it be used? If the science master is non-resident, how often is he prepared to come in again after dark? 'Tuesdays and Fridays' will not do; he must come when conditions are favourable. If he is a resident house-master, how often can he leave his boarding-house in the evenings? Handling a good telescope is a skilled job, and although from time to time there will be senior pupils who have acquired the necessary skill the science master cannot depend on them entirely. Unless satisfactory answers can be given to all these questions it is in the interests of astronomy that the gift, desirable as it is, should go elsewhere.

If the observatory is to be built, the next question is where? Conditions to be satisfied are: (a) Absence of artificial lighting and sky glare, though the latter the town school cannot escape. (b) A wide field of view: if it is not possible to cover the whole horizon, then sacrifice the north side, for the E-S-W from horizon to zenith is essential. (c) Unless the observatory is to be a very substantial one, shelter from the wind. (d) Grass surroundings rather than asphalt or concrete. These are listed in order of importance, and in addition there may be some local factor of accessibility. That of Kingston Grammar School was very superior, but timber, or asbestos sheeting, can be used instead of brick. The revolving dome can be replaced by a revolving pyramid, one sector of which is arranged to open; this is easier to make than a dome. Another arrangement is for the whole roof, which can be fairly light, to run off on rails. However, study (40) and (41), and if possible browse through old volumes of the B.A.A. *Journal*.

If the bequest is in the form of money the science master will have to decide, not only the site and type of the observatory, but what to put in it. Some time ago Dr W. H. Steavenson, who knows more about school and amateur observatories than anybody else, gave a paper on 'School Observatories' at a meeting the B.A.A. I cannot do better than quote what he said about the choice of equipment.

To my mind an ideal equipment for a school would consist of a 4-inch equatorial, with or without driving clock, together with two or three telescopes from two to three inches aperture, mounted as altazimuths on portable garden tripods. The equatorial should be mounted, on a fixed iron pillar, in a simple square box-like structure with run-off roof, allowing of an unobstructed view of the sky. The circles of the instrument will be available for instructional purposes and for finding of stars and planets in daylight, but should otherwise be used as little as possible. A driving clock is by no means essential but will be useful when objects are being shown to visitors.

The main telescope would cost new £450 to £500; second-hand ones run from £100 to £200 according to make and condition. Another important point that Dr Steavenson made is that the astronomy club should have a planned programme of objects for examination, and not just idly gaze at the same things over and over again. Such a list can be made from books such as (15), (17) and (18). This paper, and the discussion upon it, is well worth study.*

Lastly, what is the scientific value of the work done in a school observatory? I am afraid that the answer is 'very little'. Observers have to be trained, and by the time they have been trained they are likely to have left school. The science master himself may be able to use the equipment for, say, occultations, or in the case of the larger observatories the measurement of double stars, and the pupils can help, but the real aim is not achieving results now but training those who may be able to do so in the future. If they learn to find their way about the sky, to handle instruments, and to observe and record critically and honestly, the observatory has fully served its purpose.

* *J. Brit. Astr. Ass.* Vol. LXIII, No. 7 (1953).

CHAPTER V

TEACHING AIDS

Exactly what teaching aids will be needed in addition to books depends to some extent on the books available. If the class is already adequately provided with maps and pictures the supplementary material need not be very lavish, but in some schools most of such material will have to be exhibited in the class room. It is this classroom material that forms the subject of this chapter.

MAPS AND WALL CHARTS

It was mentioned on an earlier page that it would be very helpful if the pupils could have some kind of star atlas of their own, unless of course the textbook that they are using includes the necessary maps; (16) was suggested. At the same time it is useful to have a large star map for display, and there is one 46 in. × 36 in. available. This is *Philip's Chart of the Stars*, edited by E. O. Tancock, and produced in various forms, paper, cloth, mounted on rollers and so on, at prices from 7s. 6d. upwards.* For showing the sky at different times and dates, and invaluable for planning your instructions to the class on where to look for objects of interest, there is *Philip's Planisphere*. This is a revolving star map showing through an approximately circular opening representing the horizon, the middle of the opening being the zenith; it can be set for any hour of any day of the year. It is made for the latitude of London,

* All prices quoted in this chapter refer to Dec. 1961, and are liable to alteration.

but that is good enough for anywhere in the British Isles. It now costs 14s. 6d., but I feel that every school should have at least one. These two items are, of course, to be obtained through booksellers.

If there is a school telescope of 2- to 3-inch aperture a more comprehensive star atlas will be required. For many years *Norton's Star Atlas* (17) has been the amateur's standard companion at the telescope, but there is now available a simpler and cheaper version of it (18) that will meet the needs of most schools. Until you have a good telescope it is not required; when you have it certainly is.

Wall charts in the ordinary sense are scarce. There is a set of three published by Educational Productions Ltd., East Ardsley, Wakefield; 7s. 6d. each or 20s. the set.

S 671   Eclipses
S 672   Phases of the Moon
S 673   The Planets

While the stock lasts there is one obtainable from Rank Precision Industries, G.B. Film Library, Aintree Road, Perivale, Greenford, Middx.; W 23 'The Seasons', price 5s. 6d.

Wall charts and pictures for display are, of course, published in the United States. I have not seen them, but if any readers wish to make enquiries here are two addresses from the advertisement pages of *Sky and Telescope*:

> Sky Publishing Corporation,* Harvard College Observatory, Cambridge 38, Mass.
> Astronomy Charted, 33 Winfield Street, Worcester 10, Mass.

I do not know of any source in this country of inexpensive astronomical pictures for display in the classroom; presumably the demand is too small to justify publication.

---

* One of their sets of pictures is reviewed in *J. Brit. Astr. Ass.* Vol. LXV, No. 1.

Thus projection methods must be used, and the episcope (epidiascope) is one solution, enabling you to show book illustrations. In this connexion *Ball's Guide* (15) might be useful, as the maps and pictures in it are of a convenient size for this purpose.

LANTERN SLIDES

Transparencies are more effective in projection than are the opaque subjects, and there is nothing to beat the old-fashioned $3\frac{1}{4}$ in. square lantern slide. The difficulty is to get them. Slides of the highest class, made direct from observatory negatives, can be bought from the Royal Astronomical Society,* but few schools will be able to indulge in these at 7s. 6d. per slide. Slides of this character are in the loan collections of both the R.A.S. and the B.A.A., but naturally these are available only to the members of the respective societies. There is a large loan collection of $3\frac{1}{4}$ in. slides at the Science Museum, South Kensington, London, S.W. 7, fees being $1\frac{1}{2}d.$ per slide, minimum 2s. 6d., plus return postage. They can be purchased to order for 3s. each in either $3\frac{1}{4}$ or 2 × 2. Ask for list 456 part vii.

Some slides can be made by the teacher, particularly if he is a photographer, for then he can copy book illustrations at a cost of about 1s. a slide. Permission should be obtained from the publisher, and subject to the condition that the slide is for *personal use only* I have never known permission refused. The amateur slide-maker may, of course, prefer to work in the now popular 2 in. × 2 in. size.† For those without a camera there are various types of writing slide, transparent

* Enquiries to the Assistant Secretary, Royal Astronomical Society, Burlington House, London, W. 1.
† See C. H. Bailey, 'How to make Film Strips and Miniature Lantern Slides', *Sch. Sci. Rev.* No. 110 (1948).

for writing in ink, and opaque for writing with a needle, but of course their use is limited to diagrams. Constellation slides can be made quite successfully from a piece of cardboard with the stars pricked in it. I prefer to burn the holes with a hot pin, and in either case the further the pin is pushed through the brighter the star on the screen.

FILM STRIPS

By far the cheapest way of getting good pictorial material is the film strip, and astronomy is quite well served in this medium. The problem is how to use it, for the choice of the pictures and their sequence is not your own, and somebody else's film strip rarely fits exactly one's own lesson. One way out is to use the strip only as a source of pictures, and show only the two or three that are needed at the time. With an interested form it is quite difficult to get peacefully from one chosen picture to the next choice, half-a-dozen farther on! An alternative is to cut up the strip and mount what you want as 2 in. × 2 in. slides, thus giving yourself complete independence. This process is not as extravagant as it sounds; if you throw away half the strip you will still get 15–20 slides for under £1, which is considerably cheaper than 3s. a slide, or the 7s. 6d. that you would have to pay for some pictures that are now on the film strips. What I prefer to do myself is to incorporate the strip as a whole, or a longish unbroken sequence from it, into my own course. It may come at the beginning of a topic as a pictorial introduction, or perhaps at the end as a summary and conclusion. This process avoids cutting the strip and satisfies the class that they have not missed anything, and the author's sequence usually is a logical one. Some of the strips are accompanied by quite comprehensive notes. The average film strip is too long to cram into one school period, unless that particular period is simply 'we will look at some pictures to-day for a change'. A film strip lesson must allow time for questions and discussion, if not at every picture, at least at frequent intervals.

There follows a list that I prepared for the British Astronomical Association.* Strips can be bought from the respective publishers, educational suppliers and many photographic dealers.

Film strips on astronomy

The publisher's catalogue number precedes the title, and the number after the synopsis is the total number of pictures, single frame unless otherwise stated.

General astronomy in single strips

Educational Productions Ltd, East Ardsley, Wakefield, Yorks.

Gumperts 2 *Astronomy*, adapted from Swedish by E. A. Beet.
> Earth's motion—Moon—Sun—the four well-known planets —stellar objects. (36) With short notes, 17s. 6d.

Hulton Educational Publications Ltd, 55/59 Saffron Hill, London, E.C. 1.

360 *Astronomy*, by Patrick Moore.
> Telescopes—Sun—Moon—planets—comets—meteors— clusters and nebulae. (48) With notes, 12s. 6d.

A Course in three strips

Rank Precision Industries Ltd, G.B. Film Library, Aintree Road, Perivale, Greenford, Middx. With brief notes, 17s. 6d. each.

S–188 *Movements of the Sun and Earth.*
> Shape and size of Earth and Sun—rotation of Earth— revolution around the Sun. (16)

S–146 *Planets*, adapted from French by C. A. Ronan.
> Sun as a star—movements of the planets—their physical conditions. (48)

* Reprinted, with amendments, from *J. Brit. Astr. Ass.* Vol. LXIV, No. 7 (1954).

S–147 *Stars*, adapted from French by C. A. Ronan.
Observations of the stars—binaries—variables—novae—the
Galaxy—galactic and extra-galactic nebulae. (49)

A course in eight strips
Common Ground Ltd; sold by Educational Supply Association,
233 Shaftesbury Avenue, London, W.C. 2. With notes, 18s. each.

CGA: B 192 *The Earth as a Planet*, by E. O. Tancock.
Position and size in relation to Sun—latitude and longitude—
Earth's orbit—seasons—time—model of celestial sphere. (35)

CGA: B 191 *The Solar System*, by C. A. Ronan.
Sun (briefly)—positions, sizes and nature of the planets and
their satellites. (41)

CGA: B 190 *The Moon*, by E. A. Beet.
Phases—earthshine—eclipses—tides (briefly)—introduction to
surface and physical conditions. (35)

CGA: B 371 *The Sun and other Stars*, by E. A. Beet.
Size and nature of the Sun—sunspots and their effects—
chromosphere—distances of stars—giants and dwarfs—tem-
peratures—doubles—variables—spectrum (briefly). (35)

CGA: B 357 *Comets and Meteors*, by C. A. Ronan.
Structure of comets—tails—orbits—famous comets—meteors
—meteor orbits—meteorites. (39)

CGA: B 370 *The Chief Constellations*, by E. O. Tancock.
Rotation of the sky—daily and seasonal changes—visible and
invisible circles—well known constellations (including
southern)—introduction to telescopic objects. (33)

CGA: B 571 *The Stellar Universe*, by E. A. Beet.
Galactic clusters and nebulae—the Galaxy—extra-galactic
objects—the scale of space. (32)

CGA: B 732 *Developments in Astronomy 1945–1960*, by E. A. Beet.
Radio astronomy—some fields of progress—astronomy by
rocket and satellite. (36)

For geography teachers

G.B. Film Library. With brief notes, 18s. 6d. each. (S–1 also available in colour at 30s.)

S–18 *Day and Night.*
Rotating globe—rising and setting—variation in the length of the day. (25)

S–1 *Latitude and Longitude.*
A world made to open, and a transparent globe showing the internal angles. (21)

S–2 *Longitude and Time.*
Local time—G.M.T.—calculation of longitude—time zones —date line. (25)

Also SC–185, *The Earth and the Sun's Rays* (17); SC–186, *Movement Around the Sun* (22); SC–187, *The Temperate Regions* (19). Adapted from French. Colour, 30s. each.

Other strips

Visual Information Service, 12 Bridge Street, Hungerford, Berks.

98 *Great Astronomers.*
From the ancients to Adams, including Ptolemy and the Greeks—astrology—Copernicus—Galileo—Tycho Brahe—Kepler—Newton—Halley—Herschel. (72) Without notes, 10s.

99 *The Night Sky.*
Diagrammatic representations of the principal constellations and how to identify them. (36) Without notes, 7s. 6d.

259 *The Universe.*
Distance and size of Sun, Moon and stars—the Galaxy—the expanding universe. (14 double frame) Short notes. Colour, 15s.

Educational Productions

C 6437 *The Solar System.* An English version of an Italian strip. The insignificance of the solar system in space—from the Sun to Pluto as the space-traveller might see it. (31) With short notes. Colour, 30s.

Hulton Educational Publications. Two more Italian strips in English version, with short notes, colour, 21s. each.

378 *The Sun.*
Historical—distance and size—corona and chromosphere—sunspots and their effects. (29)

379 *The Moon.*
Distance and size—origin and nature—surface features—origin of craters—phases—eclipses—tides—the other side. (36)

For obvious reasons I am not in a position to comment on these strips, so here is a list of reviews that I happen to have seen:

Gumperts 2: *J. Brit. Astr. Ass.* Vol. LXII, No. 2, 1952, and *Observatory*, Vol. LXXII, No. 867, 1952.

S–188, S–146, S–147; *J. Brit. Astr. Ass.* Vol. LXV, No. 4, 1955.

CGA: B 192: *Look and Listen*, Oct. 1947.

CGA: B 191: *Monthly Filmstrip Review*, Nov. 1947.

CGA: B 190: *Look and Listen*, July 1947 and *Sch. Sci. Rev.* No. 125, 1953.

CGA: B 271: *Look and Listen*, July 1948, *Monthly Filmstrip Review*, July 1948, and *Sch. Sci. Rev.* No. 125, 1953.

CGA: B 357: *Look and Listen*, Sept. 1948.

CGA: B 571: *J. Brit. Astr. Ass.* Vol. LXV, No. 1, 1954, and *Sch. Sci. Rev.* No. 128, 1954.

Hulton 360: *Sch. Sci. Rev.* No. 145, 1960.

Hulton 378–9: *Sch. Sci. Rev.* No. 147, 1961.

FILMS

The film has its place in school, and is of value so long as it is confined to doing what cannot be done properly otherwise. In physics I use the animated diagram type of film or film loop for such topics as the propagation of waves and the induction of currents. A geography master cannot take his class to see a sheep farm in Australia, nor can a chemistry master in Bodmin show his pupils the manufacture of glass in Barnsley. The film is a solution to both these problems. But there is a danger. I remember reading in a Service manual on visual aids a remark to the effect 'don't just switch on the film and then go out for a smoke'. Of course, no good teacher would, but it does illustrate what I mean, that the film is to supplement, not supplant, the living teacher. There is much to be said for the silent or mute film, where the familiar teacher must do the talking and not hand over to the mechanical voice. The film can be such a simple way out, and I often feel that when some society says 'let's have a film' it really means that nobody is willing to take the trouble to prepare a paper. In the case of astronomy I am not at all sure that the film can do what cannot be well done without it. The charm of amateur astronomy is the individual effort and achievement that is normally associated with it: the simple way out is not the best road to achievement. There was a time when people prepared all their own food and made their own entertainment. We have long been accustomed to canned food and canned entertainment; must we now have canned astronomy? However, perhaps I am being old-fashioned; there are astronomical films and a full list has been published recently.* Here is a selection from them, all 16 mm. sound unless otherwise stated:

First we have a general interest film of the type which might be found in the programme at a news theatre.

Inquisitive Giant: running time 30 min.; distributed by the Central Film Library, Government Buildings, Bromyard Avenue, London, W. 3; hiring fee 30s.

This is a very interesting film dealing with the design and

* *J. Brit. Astr. Ass.* Vol. LXX, No. 5 (1960). Film and filmstrip list obtainable separately, price 1s., from the B.A.A.

building of the 250-foot radio telescope at Jodrell Bank, and featuring some of the men responsible.

Next we come to three groups of instructional films, the first two being obtainable from the G.B. Film Library, Aintree Road, Perivale, Greenford, Middx. Each of the Britannica series of four runs for about 11 minutes and the hiring fee is 10s. I personally should not use them as a part of a lesson, but rather to round off the course by revising from a different viewpoint. They are also suitable for showing at a society meeting.

109 UC, *Earth in Motion.*
   Shape, rotation, revolution—inclination—day and night—seasons.

148 UC, *Moon.*
   Tides—phases—apparent motion—eclipses.

118 UC, *Solar Family.*
   Planets, motions and satellites—planetesimal hypothesis—real and apparent motions—planetary details in certain cases—Halley's comet.

161 UC, *Exploring the Universe.*
   Vastness of the universe—telescopes—stars and clusters—Galaxy—expanding universe.

The second series, produced by G.B. themselves, are definitely classroom films. These are for geography teachers and cover much the same ground as the film strips from the same firm, p. 58. They are in colour and the fee is 20s.; two of them are also available in black and white at 10s. and three as mutes (i.e. without the sound track, for silent projectors) for slightly less. One of the G.B. series calls for special mention:

F 4573, *The Moon*: running time 17 min. (2 reels); hiring fee 20s. This is a very good teaching film, as indeed they all are, and serves as an alternative to film strip CGA 190.

Thirdly, suitable for the classroom or the society meeting, there are two from the Educational Foundation for Visual Aids, 33 Queen Anne Street, London, W. 1.

Earth and Sky: running time 45 min.; two reels in colour, hiring fee 30s. for each reel.

> Made in the Department of Philosophy at Leeds University and intended for both arts and science students. The first reel deals with the history of astronomy from the earliest times to about 1500, and the second covers the next two centuries.

Kepler and His Work: running time 16 min.; hiring fee 15s.
> An English version of a film originally produced in Munich and incorporating numerous contemporary prints. The title adequately describes its contents, but in addition the film fits his work into the times in which he lived.

Lastly there are the research films, consisting of speeded-up pictures of real astronomical phenomena, such as the motion of the solar prominences and the rotation of the planets. They are not teaching films, but are of great interest to those who know enough astronomy to be able to appreciate them, and the Royal Astronomical Society has several of these (silent). They are not ordinarily available to the public; enquiries can be addressed to the Assistant Secretary (see p. 54).

The British Astronomical Association can hardly be called a 'teaching aid', but teachers should know that schools can become affiliated to that body. The B.A.A. is an organization of amateur astronomers and has a membership of about 2500. Meetings are held in London, monthly during the winter, and a limited number of representatives of affiliated schools can attend meetings if they live within reach. Meetings, however, concern only a few of the world-wide membership; there is more to it than that. There is the *Journal*, issued eight times a year; the *Handbook* (45), issued annually; and other occasional publications. A small extra subscription brings *Circulars*, announcing new comets or other events requiring immediate notification. There is a lending library of books and a loan collection of lantern slides, film strips and photographs. Telescopes and other apparatus are available for loan, though newly affiliated schools cannot expect to get one straight away. Finally, there are the Directors of Sections, each a specialist in some particular field such as Jupiter or the History of Astronomy, whose advice is readily given to those who need it. All these things are available to the affiliated school, and where there is an observatory and some serious work being carried on the advantages are obvious. The standard annual subscription is forty-five shillings, and the address for further information is 303 Bath Road, Hounslow West, Middx. (postal communication only).

There are numerous local astronomical societies in the country, and the nation-wide Junior Astronomical Society, but for these the teacher must make enquiries in his own district. No list would be correct for long, and some of the secretaries' names and addresses would have changed before it could even be printed. The J.A.S. publishes a quarterly journal, *Hermes*, and organises meetings in various parts of the country; the annual subscription is ten shillings. Unfortunately, like the local societies, it has no permanent address.

Attention must also be drawn to some of the papers that may already have their place in the school library, and certainly in the town library. Among those giving forthcoming events and astronomical comment are:

> *The Times*, and the *Daily Telegraph*, on the first weekday of each month.
>
> *Nature*, in one issue per month.

Children's Newspaper, fortnightly.

The *Journal of the B.A.A.* and *Hermes* have already been mentioned. *Sky and Telescope* (Sky Publishing Corporation, address on p. 53) is a beautifully illustrated journal for amateurs, but it does cost 7 dollars a year.

Thus there is no shortage of aids for the teacher of this neglected subject. It is to be hoped that some of my readers will neglect it no longer, and that at least a few of their pupils will find the door to a lifelong and worthwhile interest.

BIBLIOGRAPHY

A selection that teachers may find useful

BOOKS ON TEACHING

(1) F. W. Westaway. *Science Teaching*. Blackie, 1929.
Contains a useful chapter on astronomy, mainly for higher forms.

(2) J. Brown. *Teaching Science in Schools*. Univ. of London Press, 1930.
Gives a course, mainly for middle forms. This book was published just as I was introducing astronomy in middle forms and has probably influenced my teaching and the present book.

(3) *Teaching of Science in Secondary Schools*, S.M.A. and I.A.A.M. John Murray, 1947.
The only reference to astronomy is a syllabus for non-scientists reprinted from *Sch. Sci. Rev.* No. 99, 1945.

(4) *The Teaching of General Science*, S.M.A. John Murray, 1950.
Includes an astronomy syllabus.

TEXTBOOKS

(5) P. F. Burns. *First Steps in Astronomy*. Ginn, 1943.
An elementary course from a mathematical standpoint.

(6) C. H. Dobinson. *Earth and Sky*. Black, 1930.
No longer available as a textbook, but teachers should read it if they can.

(6a) C. H. Dobinson. *Earth and Universe*. Longmans, 1958. Brief, illustrated and inexpensive.

(7) E. A. Beet. *A Text Book of Elementary Astronomy*. Cambridge, 1945.
An elementary course with an experimental and historical approach.

(8) E. O. Tancock. *Starting Astronomy*. Philip, 1951.
Another elementary course.

(9) H. Spencer Jones. *General Astronomy*. Edward Arnold.
Fourth ed. 1961.
The standard English textbook for students. Useful to
teachers who have any sixth-form astronomy but rather
large and expensive.

(10) E. Nightingale. *Higher Physics*, Part 1. Bell, 1948.
The chapter on gravitation is good and is intended to cover
that section of the Advanced Level Physics.

(11) Barlow and Bryan. *Mathematical Astronomy*. University
Tutorial Press, revised edition 1944.
A book for university students, but useful for teachers who
have sixth-form work.

OBSERVATIONAL BOOKS

(12) E. A. Beet. *A Guide to the Sky*. Cambridge, 1933.
A hobby book for juniors.

(13) J. B. Sidgwick. *Introducing Astronomy*. Faber, 1951.
The first half is descriptive; the second gives the constellations in turn, pointing out objects of interest.

(13a) Patrick Moore. *The Amateur Astronomer*. Lutterworth, 1957.
A more elaborate (13).

(14) Hector Macpherson. *Guide to the Stars*. Nelson, 1953.
Similar to the second half of (13).

(15) *Ball's Popular Guide to the Heavens*. Philip. Fifth ed. 1955.
A comprehensive work including an atlas of the whole sky,
monthly maps for the latitude of London, atlas of the
Moon, diagrams of the solar system etc., lists of telescopic objects, together with pictures and notes. First
written in 1892 and old-fashioned in places.

(16) *The Stars at a Glance*. Philip.
A very simple star atlas with brief notes on the constellations
and a map of the Moon. Very suitable for lower and
middle forms and costs only 3s.

(17) *Norton's Star Atlas and Reference Handbook*. Gall and Inglis.
Tenth ed. 1950.
Atlas of the whole sky covering all naked-eye stars, together
with quite extensive lists of objects for the telescope,
and notes for the observer.

(18) *New Popular Star Atlas*. Gall and Inglis. About 1950.
A simplified and much cheaper version of (17) covering the
needs of most schools.

Most of the B.A.A. publications are definitely observational,
and a few of the 'General Reading' section could, perhaps, have
been put here.

GENERAL READING

Books under this heading are legion, so the most that I can do
is to give a short graded reading list, the early books being for
juniors and the later ones for sixth forms (not graded after 35).

(19) Peter Hood. *The Sky and the Heavens*. Puffin Books.
A picture-reader for the youngest pupils.

(19a) Peter Hood. *Observing the Heavens*. Oxford, 1951.

(20) Mary Procter. *Evenings with the Stars*. Cassell, 1924.

(21) H. E. Taylor. *Wonders of the Universe*. Pitman, 1933.

(22) W. Shepherd. *The Universe*. An Eagle Book; Longacre Press,
1960.
192 large well-illustrated pages for the 12–16 age group.

(23) H. P. Wilkins. *Our Moon*. Muller, 1954.
Also useful to those who wish to make a systematic study
of the Moon with a small telescope.

(24) Sir James Jeans. *The Stars in their Courses*. Cambridge, 1931,
and Pelican Books.

(25) E. A. Beet. *The Sky and its Mysteries*. Bell, 1952.

(26) D. S. Evans. *Teach Yourself Astronomy*. English Univ. Press,
1952.

(27) H. Spencer Jones. *Life in other Worlds*. Univ. of London
Press, second ed. 1952.

(28) Edited by M. Davidson. *Everyman's Astronomy*. Dent, 1952.
A composite work by various authors, each being an authority in his own field. Useful for reference.

(29) D. O. Woodbury. *The Glass Giant of Palomar*. Heinemann, 1940.

(30) D. S. Evans. *Frontiers of Astronomy*. Sigma, 1946.

(31) W. M. Smart. *Some Famous Stars*. Longmans Green, 1950.

(32) Peter Doig. *Concise History of Astronomy*. Chapman and Hall, 1950.

(33) J. B. Sidgwick. *The Heavens Above*. Oxford, 1948.

(34) Sir James Jeans. *The Universe Around Us*. Cambridge, 1929.
The first edition gives a good idea of the state of knowledge in the 1920's.

(35) F. Hoyle. *Frontiers of Astronomy*. Heinemann, 1955.

(36) Sir Robert Ball. *The Story of the Heavens*. Cassell, 1886.
The great popular classic of the nineties.

(37) Patrick Moore. *Astronautics*. Methuen, 1960.
A reliable book on space travel; suitable for middle forms.

(37a) M. W. Ovenden. *Artificial Satellites*. Penguin Books, 1960.
Alternative to (37), for older pupils.

(38) Bonestell and Ley. *The Conquest of Space*. Sidgwick and Jackson, 1950.
An older book on space travel; of special value for its illustrations.

(39) J. Pfeiffer. *The Changing Universe*. Gollancz, 1957.
Radio astronomy for the general reader.

(39a) J. G. Crowther. *Radio Astronomy and Radar*. Methuen, 1958.
An introduction suitable for middle forms.

UNCLASSIFIED

(40) Edited by T. E. R. Phillips and W. H. Steavenson. *The Splendour of the Heavens*. 1923.
A composite work like (28), but a very large and profusely illustrated book. Should certainly be in the school library if a copy can be obtained.

BIBLIOGRAPHY

(41) J. B. Sidgwick. *Amateur Astronomer's Handbook.* Faber, 1955.
A comprehensive reference book on the amateur's equipment.

(42) J. B. Sidgwick. *Observational Astronomy for Amateurs.* Faber, 1955.
Should be in the Observational section, but is placed here as it is a companion volume to (41); it gives detailed instructions for using the equipment in the various types of observation. Too advanced for most schools.

(43) A. Matthewson. *Constructing an Astronomical Telescope.* Blackie, 1947.
An outline of the method of making a reflector.

(44) A. G. Ingalls. *Amateur Telescope Making.* Munn, U.S.A., 1946.
The standard work on this subject.

PUBLICATIONS OF THE B.A.A.

Orders to the Assistant Secretary, British Astronomical Association, 303 Bath Road, Hounslow West, Middx.

The following are likely to be of special interest to teachers:

(45) *The B.A.A. Handbook.*
The amateur astronomer's almanack, giving most of the ephemeral information needed, some of which is not readily available elsewhere. Price to non-members 9s. plus 5d. for postage.

(46) *The B.A.A. Its Nature, Aims and Methods.*
Describes the work of the various observing sections; gives a good idea of what the amateur can do, and how to set about it. Price to non-members 2s. plus 3d. for postage.

(47) F. M. Holborn. *The Beginner's Telescope.*
Choosing, testing and maintaining a telescope. Price 2s. 6d. plus 3d. for postage.

(47 a) *Careers in Astronomy.*
A pamphlet for sixth-formers who are *seriously* considering such a career. Free, but 3d. for postage.

A list of other publications can be obtained from the Assistant Secretary.

FLYING SAUCERS

This topic is not strictly relevant but is bound to turn up. There has been so much irresponsible writing about it that it is desirable that an authoritative account of the phenomena thus called should be available in the school library.

D. H. Menzel. *Flying Saucers.* Putnam, 1953.

SCHOOL SCIENCE REVIEW

As some schools will have sets of *School Science Review* I give a list of the main articles of astronomical interest.*

	No.	*Date*	
	2	1919	E. O. Tancock. Elementary Astronomy.
	3	1920	A. C. D. Crommelin. Relativity and the Recent Eclipse.
	9	1921	B. M. Neville. The Solar Eclipse of 1921.
	18	1923	G. N. Pingriff. Some Recent Star Measurements.
	20	1924	W. Garrett. A Simple Astrolabe.
	23	1925	Sir Frank Dyson. The Measurement of Stellar Distances.
(48)	24	1925	G. N. Pingriff. A Camera for Astronomical Objects.
	31	1927	H. H. Turner. The Total Eclipse of 1927.
	32	1927	G. N. Pingriff. The Astrolabe.
	39	1929	Sir Arthur Eddington. The Interior of a Star. A lecture.
	40	1929	A. W. Barton. Cepheid Variables and the Measurement of Stellar Distances.
	47	1931	J. Young. The Lunar Landscape.
(49)	53	1932	H. E. Watson. A School Astronomical Telescope.

* Mainly extracted from the *Index* compiled by D. H. J. Marchant.

55   1933   F. Hope-Jones. The Free Pendulum.

58   1933   A. Harvey. The Chemistry of the Sun.

59   1934   Sir H. Spencer Jones. The Structure of the Universe.

(50)   70   1936   W. Railston. A Simple Equatorial Mounting.

76   1938   A. Harvey. Preliminaries to the 200-inch Telescope.

80   1939   A. Harvey. The Doppler Effect in Astronomy.

106   1947   N. R. Hall. Time.

107   1947   N. R. Hall. Time Measurement.

(51)   107   1947   G. F. West. Construction and Use of an 8¼-inch Reflecting Telescope.

111   1949   E. A. Beet. Astronomy for Schools: I.

112   1949   E. A. Beet. Astronomy for Schools: II.

113   1949   G. F. West. Recent Developments in Reflecting Telescopes.

115   1950   E. J. Aiton. The Formation of Tides.

118   1951   A. Harvey. The Aurora Borealis.

(52)   121   1952   C. G. Ferguson. Radio Astronomy.

126   1954   E. J. Aiton. Irregularities in the Tides.

138   1958   J. M. Osborne. Ranging the Satellites by Doppler Shift Observations.

(53)   140   1958   R. D. Davies. Radio Astronomy.

(54)   142   1959   J. M. Osborne. The Stowe Radio Telescope.

143   1959   E. A. Beet. Notes on the Sun.

144   1960   R. O. & J. C. Kapp. Teaching How the Tides are Produced.

147   1961   F. E. Clegg. John Goodricke—Astronomer.

148   1961   E. J. Aiton. The Formation of Tides.

149   1961   A. R. Marshall. A Celestial Sphere.

Attention is also drawn to *Science Books for the School Library*, published by Murray for the Science Masters Association, where many more astronomical titles will be found.

INDEX

INDEX

Printed in the United States
By Bookmasters